SCIENCE IN THE NURSERY

A COLLECTION OF SILLY RHYMES ABOUT THE PHYSICAL WORLD

Written by
John M. Mulhern

Illustrated by
Elisha Gillette

Science in the Nursery: A Collection of Silly Rhymes About the Physical World

Produced and printed by Stillwater River Publications.

Visit our website at **www.StillwaterPress.com** for more information.

First Stillwater River Publications Edition

ISBN: 978-1-963296-28-0 (paperback) 978-1-963296-44-0 (hardcover)

Publisher's Cataloging-inPublication
Provided by Cassidy Cataloguing Services, Inc.

Names: Mulhern, John M., author. | Gillette, Elisha, illustrator.
Title: Science in the nursery : a collection of silly rhymes about the physical world / written by John M. Mulhern ; illustrated by Elisha Gillette.
Description: First Stillwater River Publications edition. | West Warwick, RI, USA : Stillwater River Publications, [2026] | Audience: Juvenile.
Identifiers: ISBN: 9781963296280 (paperback) | 9781963296440 (hardcover)
Subjects: LCSH: Science–Juvenile poetry. | Physics–Juvenile poetry. | Chemistry–Juvenile poetry. | Biology–Juvenile poetry. | CYAC: Science–Poetry. | Physics–Poetry. | Chemistry–Poetry. | Biology–Poetry. | LCGFT: Nursery rhymes.
Classification: LCC: PZ7.1.M8213 Sc 2026 | DDC: [E]–dc23

1 2 3 4 5 6 7 8 9 10
Written by John M. Mulhern.
Illustrated and designed by Elisha Gillette.
Published by Stillwater River Publications, West Warwick, RI, USA.

Dedicated with love to Susan Herzberg,
always my Chief Science Officer...

(Disclaimer: I am not a scientist.)

Keep an eye out for **bold** words — they have definitions in this book's Glossary!

Metallurgy

The Oxen & the Geese

Seven **Copper** oxen said to seven **Gold**en geese,
“Come join us as we go abroad to Naples & to Nice!”
Said the seven golden gooses to the oxen in reply,
“Our **Melting Point** is lower—if we go there we will die!
“Not so,” the oxen said to them who bore those feathers gilt.
“At worst the heat of summer’s day will cause thy bills to wilt!”
So they all set off together, both of web-bed foot & hoove,
only to discover that their **Metal** joints don’t move....

Newtonian Physics

Laws for Objects

We're told falling apples gave **Newton** the notion
of one **Law** of **Gravity**, three Laws of **Motion**.
Whether or not the fruit fell from the tree,
his ideas redefined how all physicists see.

Newton's gravity law states that objects with **Mass**es
will pull toward each other 'long their central **Axis**.
The amount of this pulling directly pertains
to how much mass each discrete object contains
& (this one's a bit tricky, as tough as I've seen 'em)
the **Inverse** of the **Square** of the distance between 'em.

Newton's *first* law of motion directly addressed
how objects in motion & objects at **Rest**
keep moving & resting, not changing their course
unless acted upon by an external **Force.**

His *second* law states that the **Accelerations**
imparted by forces to massive creations
have values defined by the strength of the force
divided by masses affected, of course.

Law number *three* (the last one in this stack)
states that when you push *some*thing, some*thing* pushes back.
This is because every physical action
produces an equal, opposing reaction.

Newton was bright. Newton was clever.
His thinking shed light on so much, so whenever
you happen to notice some motion, take pause,
& watch how all objects obey Newton's Laws.

BEAKER (XXXXL)

Chemistry

It's Everywhere!

It's there in the garden when you fertilize.
You'll find it in baking, in making dough rise.
It's present in all presents battery-powered
& there in the shampoo you used when you showered.
You can bet it's in medicine, keeping you well,
& without it, the gelatin just wouldn't gel.

Chemistry's everywhere! **Acids** & **Bases**,
Solutions & **Compounds** at work in all places!
Chemicals bone dry & chemicals soggy,
some of them see-through & many quite foggy!

Just what *is* a chemical? Will *any*thing do?
A liquid? A solid? A gas or a goo?
Yes, pretty much anything whose smallest of grains
are made up of well-known **Molecular** chains.

The science of chemistry seeks interactions,
tries combinations, studies reactions;
playing with temperatures, pressures & motions,
increasing our knowledge & finding new potions;
applying what's learned to improving the way
that chemicals help us work, travel & play.

From the numerous **Plastics** in use all around
& the glues that we use for keeping things bound,
to cosmetics we wear & vitamins swallowed
& every last recipe cooks ever followed,
it's clearer than crystal that chemistry plays
a prominent role in each one of our days.

So...

As you play through *your* day with fresh eyes, look around
& marvel at everywhere chemistry's found!

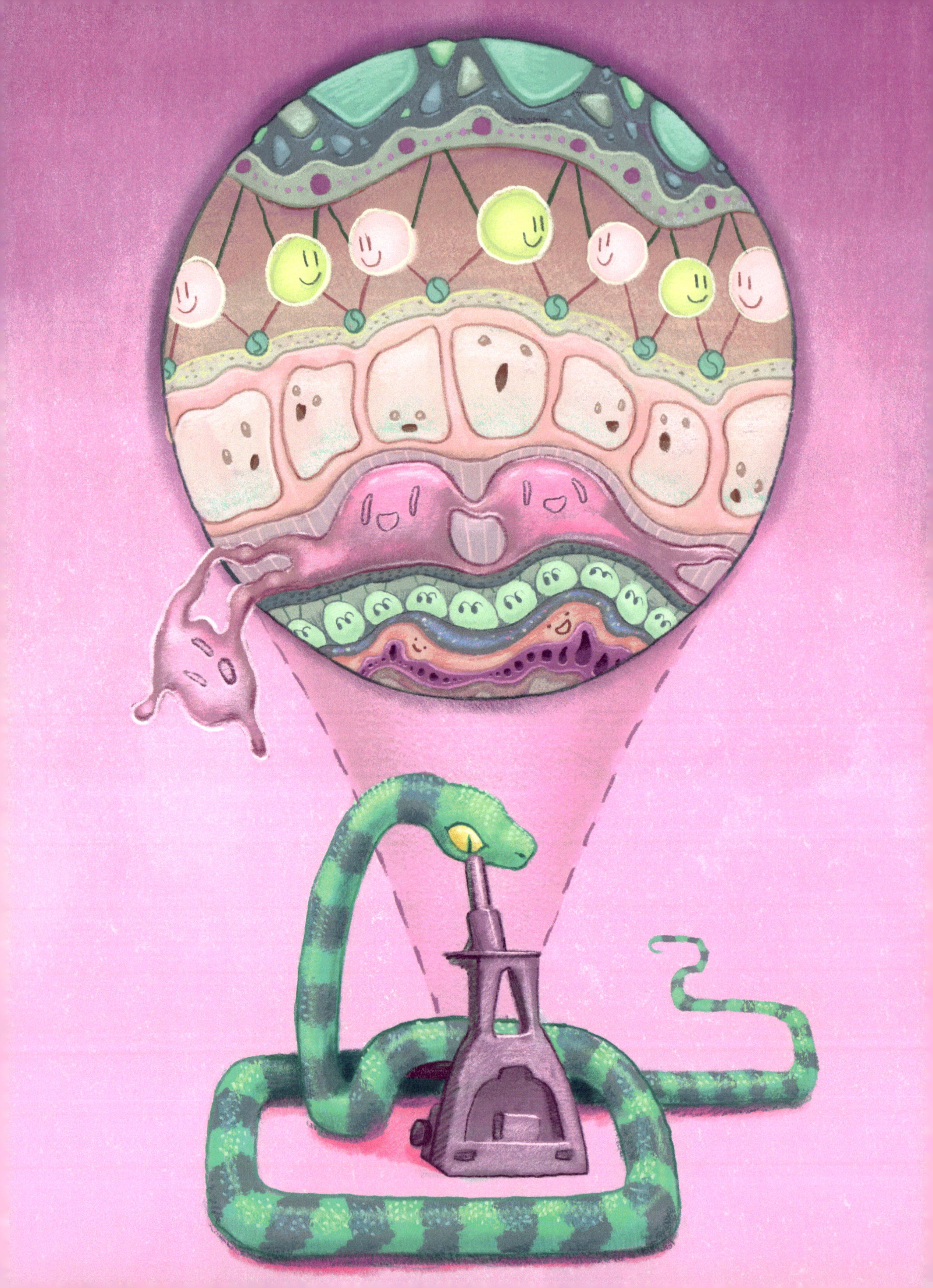

Animal Biology

System, Organ, Tissue, Cell

System, Organ, Tissue, Cell
go together very well,
forming **Bodies** when they're done
around a bony **Skeleton.**

The cells were first (& smallest too)—**Life**'s little building blocks.
They made up tissues as they grew *(not tissues from a box!)*.
These tissues make up organs that do one specific task.
Organs work in systems—where do systems work, you ask?
It's no surprise—with all they do—these wonders work inside of you!

Sending signals from our **Brains**,
pumping **Blood** through all our **Veins,**
venting excess heat through **Skin**,
fetching air that we breathe in,
breaking down the food we taste
& getting rid of yucky waste!

System, Organ, Tissue, Cell
sure work together very well.

Plant Biology

Violet & Willow

Violet & **Willow**—alike in a sense,
tho' **Flowers** are tiny & **Trees** are immense.
Still, they share many traits that **Biology** grants,
citizens both in the **Kingdom** of **Plant**s.

In spite of their jaw-dropping number of forms,
Plants tend to adhere to **Botanical** norms,
boasting similar makeups & similar ways
of keeping things **Green** in the ground or bouquets.

A great many use **Leaves** for catching the light
& **Roots** that dig deep to support veggie height!
Some send forth neat **Tendrils** for high-climbing use,
& flowers & **Fruits** are how some **Reproduce**.

In addition to **Nutrients** drawn from the land,
many plants have a food-making trick that is grand:
Making plant-food with sunlight—a deal that can't miss—
in a process that we call **Pho-to-syn-the-sis**.

This is accomplished way down in the **Cell**s
where a tiny green **Chloroplast** patiently dwells.
It soaks up the sunlight &, just like it oughta,
spins **Glucose** from **Carbon Dioxide** & **Water**!
The process of making this sugar for food
has a great side effect—our air is renewed!
For whenever plants make this nutrition appear,
they toss out pure **Oxygen** into the air.

Plant's cells, I should mention, aren't quite like our own.
The **Walls** that contain them are very well known.
Quite thick & quite tough, for protection they're king...
...& then, of course, there's the whole chloroplast thing.

So though Ol' Man Willow won't lure many bees
& violet will never be tall as the trees,
they share a strong bond, growing wild & free
in a kingdom that covers the land & the sea.

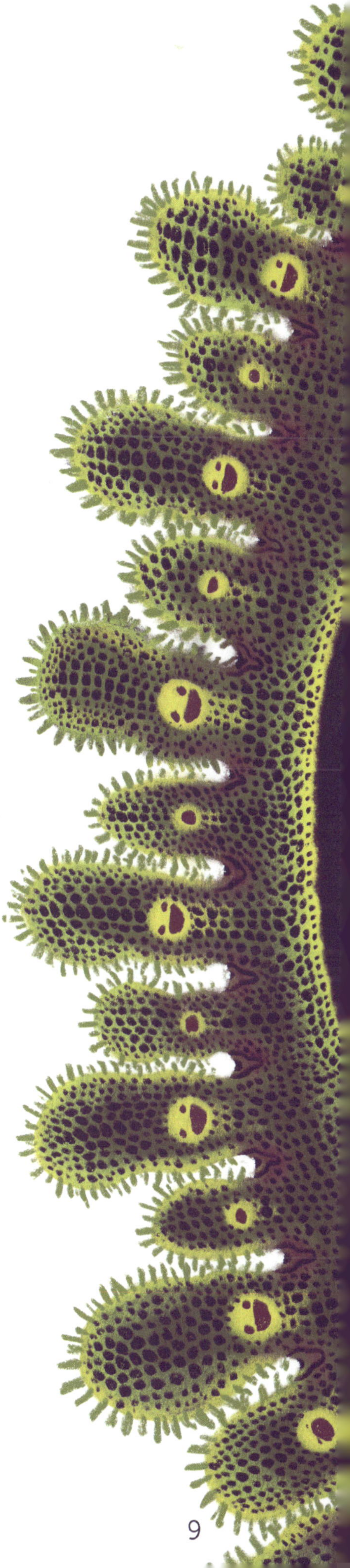

CHICKEN

Astronomy

Look Up

The daytime **Sky** is a beautiful sight,
but prettier still is the sky at night!
(One can see so much more without so much light.)
If only one *dares* to look up.

If you wish for a view that's even more pretty,
just wander away from the **Glare** of the city.
(Such human-made glows are a star-gazer's pity.)
It matters from *where* we look up.

Now, to draw in the heavens for closer inspection,
a **Telescope** is the right tool of detection.
(Don't look at the **Sun**, though, for your own protection!)
Technology helps us look up.

We look back in **Time** as we gaze into **Space**,
for the starlight we see has been on a long race.
(It shows how things were when it left its home base.)
We're *time travelers* when we look up.

The **Hubble Space Telescope**—stargazing queen!
Her **Optics** have snapped the best pictures we'd seen!
(For the poem on "optics" see page seventeen.)
From space we can *really* look up!

Astronomy's not just for **Light** that we see;
some 'scopes are outfitted for more, naturally.
(Infrared, Gamma & **X-Ray** are three.)
There are so many *ways* to look up!

Nebulas, Galaxies, Planets galore;
Pulsars & **Quasars** & **Novae** & more!
(If you find something new, you'll have reason to roar!)
It all begins when we look up!

Aeronautics

Jane's Plane

Pilot Jane had quite a **Plane** for taking to the skies in.
It loved the freedom of the air that ev'ry eagle flies in.
To start their climb, Jane took the time to add **Fuel** for the day,
which would **Combust** to create **Thrust** & send them on their way.
Her plane's **Wings** were finely shaped for cutting through the air.
They had a gift for adding **Lift** to this courageous pair!
Her airplane knew that when they flew, Jane would control their flight
by moving **Flaps** (quite hard, perhaps!) to steer them left & right.
At last, they'd come back down to earth to rest their wings...& then?
& then this team would sweetly dream of soaring up again!

Light Speed

Phillip the Photon

Phillip the **Photon**, a dashing young knave,
might be a **Particle**, might be a **Wave**.
Might not be smart but he's quick & he's bright—
Phillip the photon's a small piece of **Light**.
If he enters a race, Phillip's certain to win it
at nearly twelve million miles a minute!
Nothing is faster & he can't go slower—
our Phillip's speed limit will never be lower.
For just as the highways have signs we must heed,
the **Universe**, too, has a limit on speed.
That limit is light—it's the quickest thing going,
with photons like Phillip to keep the light glowing!

Light's speed doesn't alter, vary, or range;
light's speed is a **Constant**—it just doesn't change.
In the time that it takes you to jump off the ground,
Phillip circles the planet—*seven times 'round!*
When he's in the mood for a much longer race,
in eight minutes he'll zoom from the **Sun** to your face.
At a distance that leaves other racers in tears
is the next **Nearest Star**—it takes Phillip four years!
He's not only the fastest; he helps us to see—
so flick on the light switch & set Phillip free!

Optics

Playing with Light

What an eye-opening, brilliant delight!
There are so many ways you can play with **Light**!

Reflection, Refraction, Transmission, Dispersion,
Amplification, Deflection & Polarization.
To not mention **Scattering** would be a crime.
(It would be above but it just didn't rhyme.)

The science of optics studies the ways
that we can manipulate **Luminous Rays**.
Changing their movements, bouncing them 'round
(you'd think they'd complain but they don't make a sound!),
altering **Focus** from sharp to diffuse
& splitting them up in a **Rainbow** of **Hues!**

All of the **Colors** that all of us see
are all hidden in light that is **White**, seemingly.
Optics can tease out the violet, the red
& all colors between 'em—the whole vibrant spread!

Light's made of **Photon**s—that's all light contains.
They act like both **Waves** & like teensy bright grains.
By using some instruments built for the task,
these photons do nearly whatever we ask!

Mirrors & Prisms & Lenses & Gasses
all play well with light from the lab to eyeglasses,
helping us see & helping us know:
Wonderful science can start with a glow!

Meteorology

Whether or not
you like **Weather** or not,
an accurate **Forecast**
can help us a lot!

Will it be nice out?
Will it be storming?
What *will* today bring
on a globe that is warming?

Will **Visibility** limit our sight?
Will we find the **Humidity** suits us just right?

Should we bring an umbrella?
Or extra sun lotion?
Would a kite-flying fella
be stuck without motion?

All of these questions & more can be solved
if meteorologists just get involved.

Using tools of their trade such as **Weather Balloons**,
Gauges for rain on some wet afternoons,
all of the "...ometers" you would expect
(**Therm..., Bar...** & **Anem...** are some in this set),
computational software for **Modeling** masses
of air that we fear may be troublesome gasses,
they'll set about learning the most likely way
that weather will impact our night & our day.

While surely a science, it can't be exact;
predicting the future's a difficult act!
But meteorologists always refine
their methods & models, come rain or come shine.

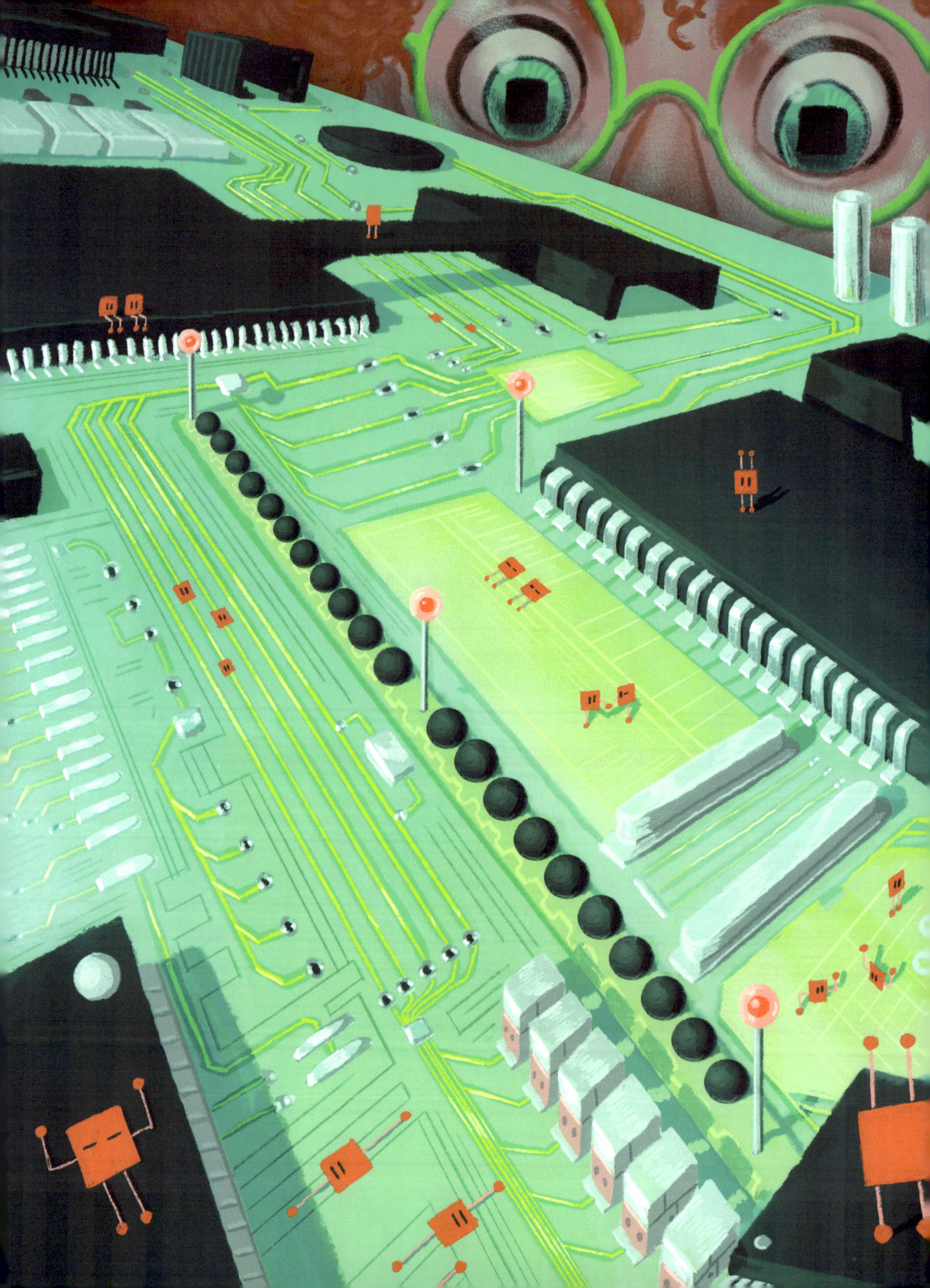

Computer Science

Bit by Bit

Computer? Science? What's to know?
Just turn it **ON** & **OFF** you go!
It's not at all a daunting matter,
Storing & **Retrieving Data,**
navigating info-spaces
with simple **GUI*** interfaces.

Memory can reach great heights
made up, it is, of **Bit**s & **Byte**s.
A **CPU** so quickly zips,
its speed is measured by its **Mips**.
Good systems boast performance chops,
their number-crunching goes by **Flops**.

For easy workflow we owe thanks
to all those in the **IT**** ranks.
They've written **Languages** & **Code,**
worked to increase memory load,
designed the **Hardware,** made it run
to bring high-tech to everyone!

There were many early heroes
in the field of **Ones & Zeroes;**
Ada Lovelace has the glamour
of, perhaps, the first **Programmer**,
& **Claude Shannon** can be quite cheery
as dad of **Information Theory**.

The work of these & many more
unlocked a fine & wondrous door,
behind which keystrokes can enable
calculations quick & stable,
custom **Programs**, data storage
& instant sharing of our knowledge!

Beneath the surfing, clicks & scrolls
reside the digital controls,
the **Operating System**, which
(provided there's no **Bug** or **Glitch**)
translates our **Input** into action,
allowing program interaction.

Then the **Software** can produce
helpful **Output** for our use;
crunching numbers, **Hyperlinking**
(listen hard—you'll hear it thinking!),
enhancing photos, searching names
& running countless nifty games!

But the next time you're computing
Online, Offline, or just **Booting,**
access your head's CPU
& retrieve this **Packet** sent to you:
computers, useful as they are,
cannot compete with all you are!

*Pronounced "gooey"
**Pronounced "eye-tee"

Engineering

Building with Science

Construction & science—
thank goodness they met!
Their marriage produced engineering!

Part **Mathematics**,
part **Physics**, part sweat.
All planning, all building, repairing.

Knowing **Materials**—their strengths & their **Stressing**—
designing machines to their last button-pressing,
Drafting buildings & bridges—whole *cities* I'm guessing!
A discipline most pioneering.

Engineers, I declares, possess very few peers
as, down through the years, they have taken great cares
to impress us to tears with their **Pulleys** & **Gears**
(not to mention their **Calipers, Slide Rules,** & **Squares!**),
leaving grateful societies cheering!

Quantum Mechanics

Wee Wally Weensy

"My, oh my," thought Wally Weensy,
"things get odd when they get teensy!"
Tumbling down a size-mic trail,
descending corridors of **Scale**,
Wally found a world so small
that "normal" hardly worked at all!
The **Laws of Nature** improvise
when shrinking past **Atomic Size**,
departing lands of certainty
for fields of **Probability**.

Wally found that if he measured
information that he treasured,
ev'ry time he gained one fact
his other knowledge would contract!
For in this tiny realm, successes
are measured by our better guesses,
& when **Observations** are agreeing
it's on more *likely* states of being.

Pick a **Particle** or **Wave**
& it is sure to misbehave,
eluding full investigation
of its **Momentum** or its station.
For qualities of place & speed
are not, when teensy, guaranteed,
existing in a cloud of fates
containing all **Potential States**.
Wally's study would, perhaps,
cause its **Waveform** to **Collapse**,
granting insight into "where"
but not the speed that got it there;
or, rather, offer confirmation
of its speed without location.

Growing back from quantum-land
with partial facts held firm in hand,
Wally smiled & said, "My word!
Nature's thrilling, if **Absurd**."

Theory of Everything

The Two Physics

Theory of everything, theory of all,
with **Rules** that work both for the **Large** & the **Small**.
A very fine theory—the best in the biz!
The problem is that...we don't know what it is.

For many long years we've known two sets of rules:
the big & the teensy—each have their own tools.
To understand "big" stuff, from dust to a star,
the **Physics** of **Newton** & **Einstein** go far.
If your interests tend to be more **Subatomic,**
then try **Quantum Physics**—the tiny-stuff tonic!

These two types of physics, though, don't work together.
Can't say that they will—neither if, when, nor whether!
A unified theory would fix things concisely
by showing how both of these physics play nicely!

This mystery theory might be about **Strings**,
might talk about **Multi-Dimensional** things.
It could go on at length about straight lines that bend
or expound upon **Membranes** that just never end.
Perhaps it will call for a new shade of green
or require new numbers like 'leventyteen.
It could, I suppose, involve super-hot **Foam**
or magnetized steam with thick hair you can comb!

It's certain to go down in all sorts of history
but for now the Grand Theory of All is a mystery.
Somebody, someday, will find the right clue.
Perhaps that bright somebody-person is you!

299792458 m/s
299792458 m/s
299792458 m/s

Special Relativity

Constant C

No theory of **Physics** can hope to compete with
our most special relatives—dead or alive—
but here's the most mind-stretching one that you'll meet with,
fathered by **Einstein** in 1905.

Building on older work by **Galileo,**
dealing with physics on sea & on shore,
Einstein said, "**Radiation,** like light, should be added,
with **Maxwell's Equations** keeping the score!"

Galileo had shown that in ships on the ocean,
objects behave as they do on the land,
whether those vessels are still or in motion—
steady speed & direction the only demand.

"Steady speed & direction" describes a key notion
in science that we call an **Inertial Frame**:
a discrete point-of-view with a uniform motion
where all **Laws of Physics** are 'xactly the same.

Relativity teaches that no frame of reference
is "better" or "worse" than another, you see.
So all points of view are quite equally valid
as long as their motion's completely change-free.

What's true for the motions of objects, thought Einstein,
must also be true for emissions like light!
From an inertial frame they would all look the same,
sharing one **Constant** *speed as they race through the night!*

This startling first relativity theory
(called "special" 'cause **Gravity** doesn't apply)
maintains that the speed of light just doesn't vary
no matter the speed of the light source or eye.

This means if you measured the speed of a sunbeam
while *you're* moving many times faster than sound,
the result you would get would be equally met
if instead you were rooted quite fast to the ground!

Because of its elegance & all of its import,
this theory became the grand toast of the age!
(For a relative theory that isn't so special,
kindly proceed to the very next page.)

General Relativity

Gravity's Grip

You've probably noticed that when you jump skyward
it doesn't take long to fall back to the ground.
It took Albert **Einstein** to finally say, "My word!
I've got to describe *why* we're all so earthbound!"

Our planet, it turns out, is **Heavy** & **Massive**
& this has an impact on deeper affairs,
for the surface of **Space-Time** is not at all passive
but bends & distorts with the **Weight**s that it bears.

Space-time is the canvas upon which we're painted,
the foundation on which the whole **Universe** stands.
All **Mass**es reshape it—they just can't escape it.
Attractions are set by what shaping demands.

A cannonball resting upon a large mattress
indents the soft surface with its heavy weight.
The **Earth** does the same in the surface of space-time,
drawing things in by the fabric's **Curved** state.

A large enough mass with tremendous **Compression**
will boast the most strength to distort & reshape.
Called a **Black Hole** it'll leave an impression
in space-time so deep even light can't escape!

So the tugging & pulling we're always observing,
this "gravity" business that holds us in place,
is simply our sliding down space-time's own curving:
a simple earth-dimple in time & in space.

Einstein soon realized that gravity's traction
felt equal to speeding up off a launch pad.
This simple idea brought him much satisfaction—
the happiest thought that he'd ever had!

energy
mass
squared
e=mc2
Einstein
speed of light

$e=mc^2$

Figure That!

If *you've* ever uttered "all things being equal,"
our friend Albert **Einstein** is partly to blame.
His **Gravity Theory**, you see, had a prequel
where **Mass**es & **Energies** both are the same.

This is expressed in a famous **Equation**
constructed by Einstein through which he declared
a thing's **Energy Value** is simply its mass
multiplied once by the **Speed of Light Squared**.

In this equation (seen up in the title),
e stands for energy, m is for mass,
c denotes light-speed (a factor most vital)
which is squared by the 2 to finish the class.

This "simple" equation drew quite a reaction
from within & without the whole **Physics** field,
bringing folks everywhere much satisfaction
to see sublime balance in nature revealed.

(Though most of us still don't quite understand it.)

Cosmology

The Big Picture

From before there was a **Universe** 'til after it is through,
is plenty time for all the things that universes do.
The study of cosmology seeks universal knowing,
to find out how it came to be & guess at where it's going.

It's birthday was a real **Big Bang**
from which all stuff around us sprang:
solid **Matter**, dark & light
(& squishy matter too, alright!).
Nuclear Forces, weak & strong,
brought **Electromagnetic Force** along.
An attractive addition, you'll agree,
was the lightweight **Force** of **Gravity**.

In fact, that huge big bang unswallowed
the building blocks of all that followed!
It's primal blasting would enable
the entire **Periodic Table**—
Carbon, Zinc &, I suppose,
the **Stuff** you find between your toes.
Ev'ry **Element** on the shelf—
yes all of this, plus **Time** itself.

Not long after this **Creation**
the whole shebang went through **Inflation**,
ballooning up, we now surmise,
to many times its former size
& then **Expanding** in a way
that continues to this very day—
speeding up as it expands
'til time at last runs out of sands.

Or, just perhaps, this universe
may one day shift into reverse,
falling backward, I've a hunch,
& ending with a real **Big Crunch**.
(That's not a very likely fate,
but still, it's fun to speculate!)

At nearly fourteen billion years our universe is old,
but billions more shall come & go before its story's told.

It's often said with great defiance:
"Bah! This isn't rocket science!"
Most speakers surely ace that quiz,
for truly, only one thing is...

Rocket Science

LIFT OFF!

N-atural learning keeps eyes skyward-turning,
A-ware that the heavens have volumes to teach.
S-o much of our Science is in close alliance,
A-cting to bring those skies into our reach.

To climb to those heights & achieve outer sights
we will need a fine **Rocket** that's up to the task.
It'll take all our knowledge (so don't forget **College**!)
& much of our science to meet what we ask.

Rocket production means lots of construction
with purified **Metals** & **Alloys** from **Ore**.
While not always light, metal's needed for flight—
being heavy, you see, doesn't mean you won't soar.

Skilled **Engineering** in each strut & bearing
ensures that our rocket is sturdy & sound.
The truth is, my dears, without skilled engineers
our adventure would never get off of the ground!

Chemistry rules when designing **Jet Fuels**,
giving our rocket the **Thrust** it requires.
The **Propellant** we mix will provide all the kicks
& will light up the sky when the **Main Engine** fires!

Newton's own **Laws** are involved in our cause,
explaining our rocket's each **Pitch, Yaw,** & **Roll**.
His insights enable a flight true & stable,
all actions & reactions kept in control.

Aeronautics, my friend, helps our rocket ascend,
carving its way through the earth's **Atmosphere**.
Its **Fins** stabilize & its **Nose** splits the skies
as it flings—without wings!—its way up through the air.

As **Einstein** declared, it is not at all weird
to see **Relativity** in our ship's path.
We'll all be relieved when our **Orbit**'s achieved
with thanks to ol' Albert for doing the math.

Bringing together this flight without feather
are **Computerized** systems onboard & on land.
All of their **Mips** make for safer space trips
(provided that **HAL** doesn't get out of hand!).

So much of our science is in close alliance,
working together to open up doors.
Learning is FUN & that fun's never done!
So get going! Get growing! The journey is *yours!*

Glossary

A

Absurd Completely ridiculous. The use of the word in the nursery rhyme alludes to a quote from American physicist Richard Feynman who, on the subject of quantum mechanics, said that it "describes Nature as absurd from the point of view of common sense. And it agrees fully with experiment. So I hope you accept Nature as She is—absurd."

Accelerations Changes in an object's momentum, brought about by the application of a force. Accelerations occur in the same direction as the force (or the aggregate direction if multiple forces are applied). Accelerations can be calculated by dividing the force by the mass affected: $F \div M = A$.

Acids Chemical compounds that 1) contain hydrogen, 2) tend to be corrosive with a sharp, sour taste (but don't taste them!), and 3) can turn blue litmus paper red.

Ada Lovelace An English mathematician born in London in 1815 who is credited with writing the first set of rules (an algorithm) intended to be read and followed by a machine (in other words, a computer program). Born Augusta Ada Byron, daughter of famed poet Lord Byron, she included her program in notes relating to Charles Babbage's analytical engine, an early forerunner of today's computers.

Aeronautics The science, based upon principles of flight, of designing, building, and flying aircraft!

Alloys Mixtures of two or more elements—of which at least one is a metal—for the purpose of forming a material boasting an improvement over the original metal, such as increased strength or corrosion resistance.

Amplification The action of intensifying light, usually accomplished by either a laser or non-thermal night-vision technology. Intense!

Anemometer An instrument for measuring wind speed. Like rain gauges, they do not work well indoors.

Animal Any living creature that can move as it wishes, has organs for sensing its environment, and a nervous system that enables a quick response to nearby changes. Often cute and fuzzy.

Animal Biology The science of animal life, sometimes called "zoology," encompassing all study of the physical makeup and life processes of animals.

Astronomy The observational science of outer space and everything in it!

Atmosphere The earth's—or any planet's—surrounding layers of gas held in place by gravity. The breathable air held within a rocket ship may also be called its atmosphere!

Atomic size The size of atoms, o'course! Among the smallest are hydrogen atoms, measuring about twenty-five picometers (which makes them about one quarter of one billionth of a meter wide). Cesium atoms, in comparison, are *huge*, measuring about 260 picometers! (You still can't see 'em, so don't strain your eyes!)

Axis An imaginary line that intersects an object either through its center of mass or through its center of rotational motion. In the case of this rhyme, we're imagining a line connecting the centers of mass of two gravitating bodies along which they converge.

B

Barometer An instrument for measuring local air pressure, useful for forecasting approaching weather systems and indicating the device's altitude, as air pressure decreases the higher up we go!

Bases Chemical compounds capable of neutralizing (effect canceling) acids, forming a salt in the process. Bases tend to be of an alkali nature (salty or, rather, salt-ish...but don't taste them!) and can restore acid reddened litmus paper back to blue.

Big Bang The leading theory on the beginning of things, mechanically speaking. This dramatic primal event—an explosion of matter and energy into a waiting void at the beginning of time—is indicated by the observed expansion of the universe; since everything's becoming increasingly spread out and cooler as time moves forward, it must have been denser and hotter in the past, making the past a bad place to store milk.

Big Crunch An imagined final victory of gravity over universal expansion. Most cosmologists now believe this to be very unlikely but that it would make a pretty good name for a brand of extra-nutty trail mix.

Biology The science and study of life and life processes.

Bit From BInary digiT, a bit is the teensiest amount of information in a binary system, usually represented as either a "0" or a "1." See *byte*.

Black Hole A term usually credited to American physicist John Wheeler, a black hole is an area of space where an amount of mass—usually the remnants of a star—has collapsed into itself to a degree that its concentration causes not merely a curve in local space-time but a bottomless "well." Predicted by Einstein's general theory of relativity and confirmed through observation, black holes command gravity fields so powerful that not even light can escape it if it gets too close!

Blood An important part of the circulatory system, blood is a mixture of several function specific cells suspended in a fluid plasma base. Categorized as a connective tissue, blood's functions are to transport oxygen and nutrients throughout the body while removing waste material, aid in the healing of damaged tissue, maintain a healthy core temperature, and defend the body from invading cells. Oxygenated blood has a bright red color that turns much darker within the body after it has delivered its cargo to oxygen hungry tissue. Bloody useful!

Bodies A *body* is the entire physical structure of an individual person or animal.

Booting Taken from the word *bootstrap*, booting is the largely automatic process of a computer becoming user-ready after it is switched "ON." This usually involves recognizing peripheral devices and loading the operating system.

Botanical Relating to botany, the study of plants (see *plant biology*).

Brains Almost certainly our niftiest organ! The brain is a very complex arrangement of nerve tissue that resides in our heads. It functions as a sort of command center for our bodies (sending the signals that control all voluntary movement and initiating many chemical functions within the body) as well as being the seat of consciousness, thought, and emotion. Pretty smart!

Bug Not a creepy-crawly in this sense, but rather an unexpected flaw that, usually due to human error, winds up in a piece of software, causing it to malfunction.

Byte A set of (usually eight) bits representing a single character of text and processed as a discrete piece of information by a computer.

C

Caliper A hinged instrument used for measuring the diameter of objects—often very tiny ones—with extreme accuraqcy.

Carbon A very common non-metallic element found in coal, diamonds, and all forms of life!

Carbon Dioxide A colorless and odorless gas, chemical formula CO_2. Carbon dioxide is breathed out by people and animals and taken in by plants to effect photosynthesis.

Cell Human, animal, and plant cells are the smallest structural part of a living thing, usually too small to be seen with the unaided eye. Human and animal cells consist of a nucleus of genetic material suspended in a watery organic gel called cytoplasm enclosed in a membrane. Plant cells differ in that they're enclosed by a thicker cell wall. Even a small houseplant can have billions of cells. About how many cells do you have in your body? About twenty trillion if you're old enough for pre-school; grown-ups have fifty trillion at least! That's a crazy lot!

Chemistry The science of substances and their properties. Seems simple, doesn't it? MWHA-HA-HA-HA-HA-HA!

Chloroplast Tiny specialized structure within the cells of many plants and algae. Chloroplasts contain the green pigment chlorophyll and carry out the process of photosynthesis to help nourish the plant.

Claude Shannon An American electrical engineer and mathematician born in Michigan in 1916 who is credited with developing information theory as it applies to electronic systems and ushering in the telecommunication age.

Code A set of instructions either performed by a computer's central processing unit during operation (machine code) or written by and for the benefit of human programmers (source code).

Collapse Referring to a particle's wave function and the effect observation has on its character. Once measured in any way, a particle must snap into some sort of regularity, shedding its "fuzziness" for something at least partly determinate, though the observer can't hope to gain equally complete knowledge of both position and momentum.

College A school of higher learning! Life itself is a constant learning experience and colleges are some of the best places for filling your mind, satisfying curiosity, or just finding out how to be a decent accountant!

Colors Perhaps the niftiest effect of visible light upon the eye! Colors are a striking chromatic effect determined by the wavelength of either the emitted light or bounced light reflecting and scattering off of objects around us before entering our eyes. Objects that absorb shorter wavelengths of light while reflecting longer ones appear redder, while objects that do the reverse tend toward violet.

Combust The controlled burning of fuel and oxygen in an engine to produce usable energy.

Compounds Chemical substances made up of two or more elements in a fixed proportion. Unlike solutions, these elements cannot be mechanically separated.

Compression If a kilogram of mass—let's say cherry cheesecake—is spread out to fill a volume of ten cubic feet, it wouldn't have a dramatic effect on the local curvature of space-time. Compress it into the space of a thimble, however, and you'd get a significantly steeper curve. Compress it still further into the volume of a DNA molecule and you've got a teensy black hole. With cherries.

Computer Any programmable device—usually electronic—designed to perform calculations, manage information, and run various programs.

Computerized Converted to computer controlled operation. The many complexities of spaceflight make the advantages of high-speed computing clear and indispensable!

Computer Science The science and study of computational theories and systems with an emphasis on practical application.

Constant A natural state or physical law that never varies. Examples of constants are the speed of light (shorthanded as *c*) and the value of the smallest allowed amount of action at the quantum level (Planck's constant or *h* for some reason). The constancy of light speed refers to its movement through a vacuum. Light does, in fact, slow down very slightly when traveling through transparent mediums of different densities. Also, if one were to set aside the "special" requirements of special relativity and measure a beam of light from the point of view of a *non*-inertial frame (a frame of reference undergoing accelerations), slightly different speeds could be recorded!

Construction The act of skillfully assembling something, usually *large*!

Copper A reddish-colored, metallic element (symbol Cu; atomic number 29; melting point 1357K, 1084°C, 1983°F). It is used widely in industry largely due to its conductivity of both electricity and heat.

Cosmology The science of the universe's structure and development—past, present, and future—mixing astronomical observation, astrophysics, and particle physics.
CPU A computer's central processing unit, responsible for calculating and executing instructions. When combined with memory, you have the fundamental core of any electronic computing system.
Creation The bringing into being of something from nothing at all, seen as an ability reserved for divinity. Many folks believe that our universe was "created," that is, brought into being by God, while others believe it happened all by itself. Discovering what you believe about the universe's origins can be a great adventure!
Curved Space-time's four-dimensional volume is difficult to imagine, but those four "surfaces" do, in fact, curve (become locally non-uniform) in the presence of mass, which in turn affects how objects move within. Both the amount of mass and the amount of space that the mass occupies effect the degree to which local space-time curves.

D

Data Plural form of *datum*. Information held in electronic form, more formally referring to a grouping of facts from which conclusions may be inferred.
Deflection The action of changing the direction of a light beam or other particle energy by bouncing it off an opaque, solid surface.
Dispersion The action of separating white light into its many colors according to their wavelength, often accomplished by a prism or moisture in the atmosphere (rainbows!).
Drafting The drawing up of detailed plans for a construction project.

E

Earth Very often a decent place to live and a marvelous vacation spot, depending on where you land. Third planet in distance from the sun (but number one in water-slides!), the earth is the home of all life that we know of. Forming, it is believed, about 4.5 billion years ago from material orbiting 92 million miles from the sun, our world today is the densest planet in the solar system thanks to a solid iron core about 1,500 miles across! With a width of 7,926 miles and a height of 7,900 miles, the earth is a teensy bit flattened because of its spin. Walking around the equator is a 24,900-mile trip, so travel light but pack a lunch. As water covers about 72 percent of the earth's surface (79 percent of the equator, so bring galoshes too), we possess over 99 percent of the solar system's liquid water (Jupiter's moon Europa likely making up the balance). Earth's atmosphere is made up mostly of nitrogen gas (79 percent) followed by oxygen (20 percent) and trace gasses. The whole business does a complete spin on its polar axis once a day and goes around the sun once a year, so even if you stay at home, you're still going places!
Einstein You don't *know?!* Well, as the uncontested rock-star of modern physics, Albert Einstein—whose name has become shorthand for "genius"*—is *the* guy. Born in Germany in 1879, Einstein would go on to redefine much of our understanding of how the world works on both large and small scales. Among his stellar contributions to science are his special and general theories of relativity, mass/energy equivalence ($E = MC^2$), explanations of both Brownian motion and the photoelectric effect, which helped establish quantum theory, and a really corking recipe for plum strudel. (This last is unconfirmed. And delicious.) Einstein's "happiest thought," referenced at the end of "General Relativity"'s rhyme, was his realization that being subject to the effect of curved space-time (gravity) was in all measurable ways equivalent to being in an accelerating reference frame and could therefore be considered exactly that! His saddest thought probably involved decelerating reference frames...or empty ice cream boxes. Yeah, empty ice cream boxes are pretty sad...

* It's also been employed as a sarcastic way of calling someone an idiot for over nine decades now!

Electromagnetic Force One of the four fundamental forces, electromagnetism deals with the inseparable relationship between electric and magnetic fields. It was nineteenth century Scottish physicist James Clerk Maxwell who first confirmed that electricity and magnetism were expressions of the same, unified force. Capable of both positive and negative charges, electromagnetic fields can combine with one another or cancel each other out!
Element Any of the 118 simplest substances we know of! Elements cannot be broken down into more basic substances but can combine to form any and all normal matter.
$E = MC^2$ The second most famous equation after $2 + 2 = 4$ but less easy to prove using jellybeans. In word form, the energy value of an object is equal to the result of its mass value multiplied by the speed of light squared (C^2). This equation represents Einstein's assertion that what we call "mass" is a measure of an object's energy content, that mass and energy are equivalent.
Energies Plural of *energy*. That kinetic quality of nature capable of causing change in the physical world, often viewed as the ability of a material system to do work or a fourth grader to clean up his or her room.
Energy Value To have mass is to have energy. All objects with mass, therefore, have an associated energy value which can be determined by running a thing's measure of mass through Einstein's little equation!
Engineering The science, craft, and art of designing, building, and operating just about any constructed thing you can think of: engines, bridges, buildings, rockets, wedding cakes, and all of Lady Gaga's wardrobe!
Equation A formulated statement of equivalence (sameness) between two things or sets of things written out with an equal sign (=) between them.
Expanding Following the very brief but energetic inflationary period just after the big bang, the universe continued to increase in volume and, in fact, has been speeding up! This expansion is universal,

meaning there's no "center" that things are moving away from, but rather all points in space are moving away from all other points in space. This is only observable at large scales, which is why you don't notice your brother's nostrils getting further apart.

F

Fins Flat projections attached to the outside of a rocket designed for smoothly slicing through the air and stabilizing the vehicle's flight.

Flaps A non-technical term for ailerons, elevators, and rudders. These are special movable surfaces on an aircraft's wings and tail which allow the pilot to alter the direction, attitude, and speed (when braking) of the craft by temporarily increasing drag on those areas. Birds' wings don't have flaps, but they flap 'em like crazy to make up for that.

FLOPS Stands for FLoating-point Operations Per Second. This is a measure of a computer's performance in dealing with floating-point calculations. What are they? Let's just say floating points are a method for simplifying number values and leave it at that (more complete descriptions of floating-point method tend to make my head hurt somethin' fierce!).

Flowers The seed bearing part of some plants containing reproductive organs which are typically surrounded by colorful petals. Often given as gifts to celebrate, say, an anniversary (or to apologize for forgetting one).

Foam At the absolute teensiest basement scale of reality, the Planck scale (way down at about 10^{-35} of a meter—many, many times smaller than a proton), reality may take the form of a chaotic, highly energetic "super-foam" where virtual particles are constantly being created and destroyed and where the laws of general relativity finally unite with quantum mechanics! (And you thought a whole lot of nothing was happening under your fingernails!)

Focus The amount of definition a beam of light has, depending on how confined the direction of its photons are. Highly focused light tends to occupy a narrower space along its direction of travel and creates a sharper image on any surface it strikes.

Force[1] (Newtonian Physics) Any energetic interaction that changes the momentum of an object. When applied, the object will accelerate in the direction of the force. The greater the mass of an object, the more force is required to produce a set amount of acceleration. A force's value can be calculated by multiplying the object's mass by the acceleration experienced: $M \times A = F$. (If you were told there would be no math, I do apologize.)

Force[2] (Cosmology) Here, referring to the four fundamental forces—or "interactions"—that underlie all changes we observe in creation. Gravity, electromagnetism, strong nuclear force, and weak nuclear force serve as our engines of both persistence and change. May they be with you.

Forecast An educated estimation of the likelihood of near-term weather patterns.

Fruits The fleshy, seed-containing growths of most flowering plants. Most fruits have a sweet taste and are easily accessible to humans and animals, making them an excellent way for trees and other plants to spread their seeds far and wide!

Fuel Chemicals that, when burned in an engine with an oxidizer (usually oxygen), produce energy to propel automobiles, locomotives, airplanes, rockets, and probably flying saucers!

G

Galaxies Massively huge collections of stars and other matter held together by gravity. Our own galaxy—the Milky Way—contains as many as 400 billion stars (some smaller than our sun, some much, *much* bigger) and stretches over 100,000 light-years across!

Galileo Called the father of modern science, Galileo Galilei was born in Italy in 1564 ("modern" being itself relative). Growing into a first-class astronomer, mathematician, and physicist, he would harness the telescope as the most important tool of sky-watchers everywhere, having greatly improved upon existing models. Also, by determining the unchanging nature of physical laws under conditions of steady motion, he established the first framework of relativity.

Gamma The most energetic form of light! Dwelling far above the visible range, gamma radiation boasts the tiniest wavelength. Exposure to large amounts of it will make you sick rather than turn you into a superhero. Sorry.

Gasses The third state of matter after solid and liquid. Gasses are characterized by low density, completely free molecular motion and the tendency to expand into any available space. Since light moves at slightly different speeds through different mediums and materials, gasses of different densities can alter the speed—and therefore the direction—of light. Our own gaseous atmosphere can alter light in remarkable ways, producing a number of interesting and beautiful effects such as mirages and sunsets. I wish I could end with a funny comment or joke about gas, but there just aren't any....

Gauges These are simple devices for measuring the amount of rainfall produced by a weather event. Consisting of simple tubes with flat bottoms and graduated marks in millimeters or inches on the side, they are among the oldest weather instruments—and the only ones you can drink out of after recording the results!

Gear A specially shaped component of a mechanical assembly—secured at a single point around which it may revolve and fitted tightly with other moving parts—that transmits motion through rotational action.

General Relativity Einstein's broader follow-up to special relativity, this theory deals with accelerations and gravity. Presented to the world in 1915 (ten full years after his special version), this theory shocked the scientific world with an entirely new description of how gravitation works. Rather than masses exerting a direct pull on each other, it is now understood that the presence of mass deforms the space-time in which it's embedded, and these deformations, in turn, define how those masses must move. If one places two bowling balls on a mattress about a foot apart, they quickly come together. This is because their weights cause the mattress to curve, and that curving causes the balls to roll into each other. It seems simple, but imagine that the mattress was invisible. All you would observe is the bowling balls moving together, and you might conclude that they were somehow pulling on each other. Space-time is invisible and physicists from Newton until Einstein made a similar logical assumption about objects and mass. General relativity also predicted phenomena such as time dilation, gravitational lensing, and black holes, which have since been confirmed through testing and observation.

Glare Refers here to the aptly named problem of light pollution. This bane of sky-watchers is caused by human-made artificial lighting that rises up from cities into the night sky, making many stars and planets unobservable. Grrrr...

Glitch Similar in effect to a bug, a glitch is a momentary, often non-repeating malfunction of a computer or other electronic system. Unlike bugs, glitches often seem to fix themselves!

Glucose A simple form of sugar that serves as an important source of energy for both plants and animals. Glucose is the product of photosynthesis in plants.

Gold A yellow colored, precious metallic element (symbol: Au; atomic number 79; melting point 1337.5 K, 1064°C, 1948°F) known for its heavy weight, great beauty, and numerous uses in industry and decoration. Can be polished to a very shiny finish. *Golden* can either mean "made of gold" or "gold colored."

Gravity The weakest of the four fundamental interactions (formerly "forces") at work in the universe, observable as the apparent attraction of masses to each other, such as the pull of the earth's mass on most of our stuff, with the exception of our helium balloons (told you it was weak!). Newton understood this to be a "force"—that is, a fundamental act of nature whose influence reached out instantaneously across empty space, attracting masses and keeping planetary bodies in orbit. Contemporary understanding holds that the masses don't so much attract each other as they cause local space-time to curve, altering the movements of other masses. Gravity, while being the weakest of the four interactions (the other three being electromagnetism, strong nuclear force, and weak nuclear force), nevertheless boasts some unique qualities: it cannot be shielded against, it operates over infinite range, and it is active on all objects and particles with mass. I *shall* say it: *heavy*!

Gravity Theory Einstein's theory of general relativity (see the rhyme on page thirty-one!).

Green A bluish yellow or a yellowish blue—your choice. In plants, a green color is a result of the presence of chlorophyll, a chemical vital to the organism's ability to make food from sunlight.

GUI Stands for *graphical user interface*. A GUI is any visual method for interacting with an electronic device, utilizing graphics and menus instead of text commands. Both Microsoft Windows and Apple's Mac Operating System have been using GUIs, like, forever!

H

HAL Short for *Heuristically programmed ALgorithmic logic circuit*, the HAL-9000 computer is a fictional, and rather troubled, character invented by Arthur C. Clarke and Stanley Kubrick for their landmark novel and film collaboration *2001: A Space Odyssey*. HAL...well, he did some bad things. But it was all due to human error....Really.

Hardware The physical, touchable parts of a computer, including things like disc-drives, circuit boards, and wiring.

Heavy In every sense a relative term, *heavy* indicates a greater-than-average gravitational effect on an object. Speaking in regards to relativity, the more mass an object has—and the more concentrated that mass is by volume—the more dramatic an effect it will produce in local space-time, placing the object in a steeper curvature so that its inertia (resistance to movement by a force) is increased. So it's, like, heavy. The earth? It's estimated to weigh in at around 6,584,800,000,000,000,000,000 tons. Which is why no one steals it.

Hubble Space Telescope One of the grandest scientific instruments ever built, the HST is a bus-sized (forty-three-foot-long) reflecting telescope that was placed into Earth's orbit in 1990 at an altitude of about 350 miles up, which frees it from the distorting effects of weather and atmosphere. Named after the famed twentieth century astronomer Edwin Hubble, the HST contains instruments for exploring the sky from the near ultraviolet through the near infrared, including its nearly eight-foot diameter visible-light-gathering mirror. Hubble has produced some of the most information rich—and beautiful—visible-light images of our universe that have ever been recorded!

Hues Varieties of a given color defined by their strongest wavelength. Periwinkle is a hue of blue...I think. It might be a hue of violet...or maybe some kind of elf or pixie...'cause who would give an ogre or troll a name like Periwinkle, am I right?

Humidity The amount of moisture in the atmosphere. High humidity makes it seem hotter than it actually is by interrupting our bodies' cooling system—sweat! When sweat evaporates off our skin into the air it takes heat away with it, but when the air is already saturated with moisture, the sweat has no place to go!

Hyperlinking A computer software function used extensively online that allows a reader to quickly access separate pages of related information by clicking on highlighted words or pictures. Such words and pictures are called *hypertext*.

I

Inertial Frame Any frame of reference that is at rest with itself; that is, undergoing no accelerations or changes in motion. From such a laboratory, all natural laws—and experiments to test them—behave identically regardless of one's speed relative to other frames of reference.

Inflation In cosmology, inflation theory states that just after the big bang, space-time itself—the four-dimensional volume in which the universe rests—went through a brief period of hyper-fast expansion, going from about a quadrillionth the size of an atom to about the size of a marble in the tiniest fraction of a second! Following this, expansion of both space-time and all the stuff within it continued but at a slower rate.

Information Theory A set of working ideas involving electrical engineering and applied mathematics that deal with the coding of information as well as the limits of data processing, compression, storage, and transmission.

Infrared A kind of light that is below the visible range. Infrared light occupies the space on the electromagnetic spectrum between visible red and radio waves.

Input Describes any information or command that a person gives a computer. Keyboards, mice, and styluses are all examples of input devices. See *output*.

Inverse The opposite or reverse of a known value.

IT Stands for *information technology* and refers to the professional study and practical application of computerized systems to store, retrieve, process, and transmit digitized information as well as fixing any related problems that crop up along the way!

J

Jet A reaction engine that produces thrust by burning fuel and oxygen, channeling the explosive result in one direction so as to accelerate itself in the opposite. Aircraft jet engines take oxygen in as they fly while the rocket variety carries its oxygen along with the fuel.

K

Kingdom One of the highest (in older traditions, the very highest) classifying rank of living creatures. The two oldest and most noted kingdoms are the plant kingdom (which includes all plants) and the animal kingdom (containing all animals). Within each kingdom, creatures are further categorized into increasingly specific groups of related life-forms. Traditional descending order of categories are: kingdom, phylum, class, order, family, genus, species.

L

Languages Refers to programming language—any precise system of symbols designed to relate instructions to a computer. All computer programs are written in one or another of these. Some examples of programming languages are Fortran, COBOL, C, C++, Java, and Perl.

Large The universe as experienced on macroscopic scales (though this still covers some pretty small stuff). Generally speaking, we're talking anything larger than a molecule. Objects that inhabit these scales tend to behave in pretty common sense ways that follow the predictions of Newton and Einstein.

Law In the realm of the sciences, laws are universally accepted conclusions, drawn from observation and testing, on how the natural world works under given conditions. As these laws are, by their nature, unbreakable, no arrests have been made.

Laws of Nature Commonly held, invariable regularities in the workings of the natural world, usually occupying a prominent role in scientific practice. At tiny scales, some of these regularities break down while others simply become irrelevant.

Laws of Physics As above, these are universally accepted and unalterable conclusions—backed by observation and testing—on how masses, forces, electricity, magnetism, and particle qualities behave.

Leaves The primary organ of photosynthesis—the food-making ability of most plants. Leaves tend to be thin, flat structures well-suited to catching the sun and can vary in size from smaller than your fingernail to larger than a dinner plate!

Lenses Optical tools made of transparent crystal or glass, usually disc shaped with one or both sides curved, used for either focusing, or dispersing beams of light.

Life The energetic state that bricks, dryer lint, and stepped-on ants do not have. Generally, something is said to have life if it can metabolize food (eat), grow, reproduce, and die. (That last one is considered a dead giveaway.)

Lift The upward force on an aircraft produced by the action of its wings as they move through the air. Lift is one of the four forces acting on an aircraft in flight. The other three forces are gravity (which technically isn't a force at all but rather a consequence of curved space-time!), thrust, and drag. To keep the craft up in the air, lift must be stronger than the gravity pulling it down. To keep the craft moving forward, thrust must be stronger than the drag (air resistance) opposing its motion.

Light Broadly speaking, light can refer to all forms of electromagnetic radiation not just the narrow band that our eyes can detect. Telescopes can be constructed to see all of the stuff that we miss! This can be especially helpful when regions of space are obscured by dust and gas that block visible light but not other forms. *Visible light* is that part of the electromagnetic spectrum that makes the sense of sight possible. By bouncing and scattering off of physical surfaces before entering our eyes, it allows us to form an awareness of our environment and the objects within it. Light is made up of massless, chargeless particles called *photons* that usually behave like waves. No one's sure why. Optical tools such as lenses and mirrors can cause light to behave in different ways that can increase our understanding of light and how we see. At a speed of 670,616,629 miles per hour, light would do serious damage to everything it touched if only it had mass. Which it doesn't, so don't be afraid to greet the dawn.

Light Speed As fast as fast gets! Light travels at 983,571,056 feet per second, which is also 299,792,458 meters per second, which comes out to 186,282 miles per second *or* 670,616,629 miles per hour.

Luminous Rays Well-defined beams of visible light.

M

Main Engine While a rocket ship may have several engines, the engine of primary thrust enjoys this distinctive title.

Mass The quality of matter that resists change in motion (inertial mass) and effects an apparent attraction to other matter (gravitational mass, often thought of as *weight*). Both inertial and gravitational mass are aspects of the same quality, having identical value in any given object. Unlike true weight, which changes depending on the strength of local gravity (you weigh much less on the moon than in, say, Scranton, Pennsylvania), mass is an intrinsic, material quality of an object that isn't altered by gravity. Relativity understands mass to be a measure of an object's total energy content.

Massive Technically, something can be said to be *massive* if it has any mass at all. The word is usually used, however, to describe something with an impressive amount of the stuff.

Materials The physical matter used in construction. Engineers need to know a great deal about the materials to be used in a project so that all of the parts perform properly and meet whatever demands are placed upon them. (The little pig who built his house out of straw was not a very good engineer.)

Mathematics The study of quantities and magnitudes as expressed through numbers and calculations. *Math*—as the lazy refer to it—may be studied in its own right (pure mathematics) or as it applies to practical use, such as in engineering (applied mathematics).

Matter All the stuff that isn't energy. Normal matter emits, absorbs, or reflects energy, enabling us to detect it. Oddly, this seems to account for only about a quarter of all matter. The other three quarters or so is called *dark matter*, as it neither emits, absorbs, or reflects energy and is only detectable by its observed gravitational effect on normal matter.

Maxwell James Clerk Maxwell was a Scottish physicist. Born in 1831, he is famous for revolutionizing our understanding of electricity and magnetism as a single force through his groundbreaking equations. He is also credited with discovering that light is itself a form of electromagnetic radiation.

Maxwell's Equations A set of four famous equations composed by Scottish physicist James Clerk Maxwell that describe the properties of electromagnetic fields. By incorporating the consequences of these equations into classic Galilean relativity, Albert Einstein formulated his theory of special relativity.

Melting Point The temperature at which a given material is hot enough to pass from a solid to a liquid state (nominally at sea-level air pressure). Melting points vary widely between materials. Tungsten? 6192°F. Chocolate? About 88°F. Mmmmm...

Membranes In some models of creation, everything within a given perceivable reality—including its own space-time—is embedded within a slice of a greater reality called a *bulk*. These slices are called *membranes*, or "branes" for short, and our own universe's membrane, if true, might well be one of countless parallel membranes. Our own local big bang may have been the result of an extra-dimensional collision (or maybe just a bump) with a neighboring membrane. Thanks, neighbor!

Memory The place on a computer where stored information resides. The two main types of computer memory are random access memory (RAM) and read-only memory (ROM). RAM is designed for rapid access and contains data that can be easily modified, added to, and deleted. ROM, on the other hand, is meant for permanent storage and takes longer to access.

Metal Any of a variety of elements found in the earth that tend to be hard and shiny, are good conductors of heat and electricity, and can be deformed without shattering.

Metallurgy The science of metals dealing with their extraction from metallic ores found in the earth, their refinement, engineering applications, and performance.

Meteorology The science of atmospheric processes and changes with a focus on foreseeing near-term weather patterns.

MIPS A measure of computer processing speed. Stands for *millions of instructions per second*. Put backwards it becomes "SPIM," which may stand for *science poetry is mild* (when compared to many limericks, anyhow).

Mirrors Not just a help when combing your hair, mirrors are optical tools that can bounce light back while maintaining nearly all of its energy and character. Non-flat mirrors (called *concave* when they curve inward, *convex* when they bulge out) can distort light waves and the visual information they carry, making images of objects reflected look compressed or bloated!

Modeling A weather model is a computer algorithm (program) that takes real-time information from various sources (such as satellites, radar, and ground instruments) and processes it all into a detailed prediction of what the weather will likely be doing over time, from sea-level to the upper atmosphere. Only computers are fast enough to do this, as by the time humans crunched all the numbers our forecast would be for six months ago!

Molecular All matter is made up of molecules—very tiny groups of even tinier atoms (you can't see 'em!). The type of atoms in a molecule and their number define the properties of the substance they form.

Momentum Take the mass of a moving body, multiply that by its velocity, and you've got momentum! The more momentum something has, the more energy it takes to slow it down, or the more energy is released if you stop it suddenly, like with a brick wall.

Motion The process of changing position over time relative to the position of other objects or frames of reference. Depending on an observer's own position and state of motion, the motion of other objects may seem to be accelerating (speeding up), decelerating (slowing down), or steady. It's important to note that no motion is absolute in and of itself but exists only in relation to other objects in different states of motion.

Multi-Dimensional Occupying more than one dimension. Our everyday experience is with three spatial dimensions (length, width, and height) and one temporal dimension (the one-way passage of time). Many more dimensions may exist, though confirming their existence is a daunting challenge.

N

NASA Founded in 1958 under President Eisenhower, the National Aeronautics and Space Administration manages America's civilian space program. The talented people at NASA have been the caretakers of our stellar dreams for over half a century, helping to lift us into orbit and to the moon as well as placing roving eyes and laboratories on distant worlds! Establishing it as a civilian agency fostered the ideal of space exploration as a peaceful endeavor for the benefit of all humankind.

Nearest Star Proxima Centauri is the nearest extra-solar star. A relatively cool, dim red dwarf (like most of the stars in our galaxy), Proxima Centauri lies about four light years and three light-months away in the Centaurus constellation. That's 265,612 *times* our distance to the sun, or about twenty-four trillion miles. Due to the slow burning of its fuel, it is estimated that Proxima Centauri will continue to burn for another *four trillion* years! By then, our own sun will have been dark and dead for well over three trillion years! It pays to keep cool.

Nebulae Big clouds of gas and dust in space. Nebulae may sometimes hamper astronomical viewing but can often be quite beautiful if lit and photographed well. Like your sister.

Newton Sir Isaac Newton, Englishman (1642-1727). One of the very greatest scientists and mathematicians of all time! Newton's enormous influence on physics and optics as well as mathematics will continue to be felt long after this book—and anyone holding it—is dust! In addition to formulating his famous laws of gravitation and mechanical motion (partly adapted from the work of Galileo), he also demonstrated that white light is made up of many colors, perfected the first practical reflecting telescope, and is considered one of the fathers of calculus! (Note: this represents the first time an exclamation point has ever appeared after the word *calculus.)*

Newtonian Physics Sometimes called *classical mechanics*, this is the science of commonly observable interactions that play out between the masses and energies of objects ranging in size from grains of sand to galactic clusters. Set down in large part by Galileo, this science was further defined and codified by Isaac Newton with his law of gravity and three laws of motion. Its counterpart is quantum mechanics which deals with state changes at deeply microscopic scales.

Newton's Laws Isaac Newton's grand gift to physics—his law of gravity and three laws of motion.

Nose The tippity-front of a rocket or aircraft, usually tapered to a narrow point to decrease air resistance during atmospheric operation. Picking a nose is very important when designing a rocket!

Novae The plural form of *nova*. A nova occurs when a star—usually a white dwarf in a two-star system—accumulates a layer of compressed hydrogen on its surface from its companion star. Under the right conditions, this dense, superheated hydrogen will ignite in a runaway fusion reaction, causing the star's brightness to suddenly increase many thousands of times before finally fading back to its original level. This can cause nearby space roosters to crow prematurely, waking space farmers up way before morning.

Nuclear Forces Two of the four fundamental interactions at play in the universe (the others being gravity and electromagnetism). The weak nuclear force makes radioactivity possible, while its stronger sister holds together the protons and neutrons of atomic nuclei. Both nuclear forces are potent only over very tiny distances.

Nutrients Any substances that nourish life—you probably know them as *food.*

O

Observations To observe—to study visually—is perhaps the most passive method for gaining knowledge and insight about something. When that "something" is a sub-atomic particle, however, even this activity affects the information we gain. To observe a particle means that at least one photon of energy has either bounced off of it or been emitted from it before entering our instrumentation. This very collision/emission must change the momentum and position of the particle going forward; to what degree inevitably entails guesswork!

OFF The other, less exciting state. (See "ON".) Indicates that a device is powered down, nonresponsive, and useful only as perhaps an awkward weapon or as a means of keeping loose papers from blowing away.

Offline A state of disconnection with another computer network such as the internet. This will cause your computer to haz a sad. :-(

ON The more exciting of the two fundamental states of any electronic device (a third state—*sleeping*—only applies to some). Describes when a device is energized and ready to do work!

Ones and Zeroes Refers to binary computer code, which use a two-numeral system—combinations of ones and zeroes—to encode data.

Online The state of being connected to either the internet or to some other electronic network with the ability to send or receive data.

Operating System Sometimes called the OS, this is the fundamental and necessary collection of software that allows a computer to function and run programs.

Optics[1] (Optics) The science of sight and the actions of light (bonus rhyme!).

Optics[2] (Astronomy) Here, referring to the light collecting elements of the Hubble Space Telescope, most notably its amazing mirror and the software that helps it see!

Orbit The curved motion of one gravitating body about another. When our rocket ship achieves orbit around the earth, it is essentially in a sustained state of falling towards the planet's center of gravity while constantly missing the surface, since the curve of its fall matches the curve of the earth.

Ore Any naturally occurring rock or mineral deposit which contains metals or other valuable elements.

Organ A highly structured grouping of tissues organized to perform a specialized task. Eyes, lungs, leaves, and roots are all examples of organs.

Output Describes any human readable information that has been processed by a computer and displayed in some fashion. Screens, printers, and even speakers are examples of output devices. See *input*.

Oxygen A colorless and odorless gas—chemical formula: O—necessary for animal life. Oxygen makes up about one fifth of the air we breathe and is produced by plants as a by-product of photosynthesis.

P

Packet A block of formatted data designed to be easily shared across a computer network. Or a pouch of delicious cocoa designed to be easily shared among thirsty friends. Mmmm...

Particle An itty-bitty piece of matter, smaller than an atom, that may have qualities such as mass, charge, spin, etc.

Periodic Table The periodic table of elements is a nifty chart that organizes all known elements (most basic substances) into related groups.

Photon The very smallest bit that light is made up of. Photons are massless, chargeless particles that exhibit wave-like qualities when studied. Or massless, chargeless waves with particle-like qualities, if you prefer! (Light always acts as if it's made out of waves. Which it is. But if you go looking for particle behavior in light, you'll find that too. Light just tries to please everybody!) Photons travel at the speed of light and serve as the force carrier for light and all other electromagnetic exchanges. So they're *very* busy and often can't get away to the gym for even thirty minutes!

Photosynthesis The process by which green plants make a simple form of food by using energy from sunlight to split water into hydrogen (which is retained) and oxygen (which is released). The hydrogen is then combined with stored carbon dioxide to produce a sugar called glucose. Yum!

Physics The science and study of the material and energetic world focusing on interactions between masses and forces as well as electricity and magnetism. Such study becomes very different at very tiny scales.

Pilot One who has been trained to fly an aircraft. A pilot is responsible for maintaining control of an aircraft's speed and direction, both in the air and on the ground.

Pitch Any up and down rotational movement of a flying vehicle (nose goes down while tail goes up or vice versa).

Plane Short for *airplane*—a heavier-than-air (so, not a balloon or blimp) powered (no gliders either) vehicle of the skies!

Planets Naturally occurring spherical objects in space that directly* orbit our sun or another star while keeping very close to the stellar plane, are large and massive enough to keep their orbit clear of debris, and shine by reflected light.

Plant Biology The science of plant life, sometimes called *botany*, encompassing all study of the physical makeup and life processes of plants.

Plants Living organisms typically characterized by cellular walls, an absence of sensory organs, a lack of mobility, no fixed predictable body plan, the ability to absorb water and minerals through an underground root system, and the further ability to utilize sunlight to manufacture nutrients. Wow! Really makes you respect a salad!

Plastics Human-made compounds that can be shaped by heat and pressure into countless useful forms such as bottles, toys, and even textiles for clothing.

Polarization The action of limiting the vibrations of an electromagnetic field to fewer planes of oscillation, resulting—in the case of light—in a visibly altered character. Light can become polarized to varying degrees through reflection, refraction, transmission, or scattering.

Potential States The Copenhagen interpretation of quantum mechanics—developed chiefly by Niels Bohr and Werner Heisenberg—states that at subatomic levels, a particle exists simultaneously at a range of likely positions and holds a range of likely momentums until an act of observation forces it to "decide" where it is and what it's doing. This is, of course, ridiculous. And probably true.

Prisms Optical tools made of transparent crystal or glass, usually triangular in shape with flat refracting surfaces, that can split a beam of white light into its rainbow of colors. Also employed in the 1970s as a memorable motif in the music industry.

Probability Informed estimation of the likelihood of a possible event or state of being.

Programmer Someone who writes programs for computers. I know that you know that I know that you knew that already!

Programs Any set of rules and instructions written in programming language to be carried out by a computer. Also called *software*.

Propellant Any combination of chemical fuel and oxidizer which react to produce thrust.

Pulley One of the earliest machines! A pulley is a simple device used for lifting a heavy load. In its most basic form, it consists of a freely rotating suspended wheel with a groove to hold a rope, belt, or chain. Pulling down the free end of the tether raises a load secured to the other end.

Pulsars Rapidly spinning neutron stars—very small and very, very dense—that emit highly regular pulses of radiation, most notably radio waves. Pulsars actually emit their radiation steadily, not in pulses, but—much like a lighthouse—their rotation makes their

---

* Moons directly orbit their planet and only indirectly orbit the sun

emissions visible to us at very regular intervals.

Q

Quantum In quantum theory, a *quantum* represents the smallest building block of a quality (angular momentum, action, etc.) that cannot be reduced to smaller amounts. This goes against the previous view of a smooth universe where the value of such things could be reduced indefinitely.

Quantum Mechanics The science of itsy-bitsy interactions as informed by quantum theory, the idea that all actions and momentums are made up of indivisible set values—*quantized*—at the tiniest levels.

Quantum Physics Also called *quantum mechanics*, this physics attempts to describe the behavior of objects and energies at extremely small scales. It is based upon quantum theory, which states that amounts of energy, momentum, and even gravity are each made up of incredibly small, uniformly-valued building blocks that cannot be subdivided into smaller units.

Quasars Short for *quasi-stellar radio source*, quasars are the very energetic cores of some extremely distant—and rapidly receding—galaxies. It is believed that super-massive black holes in the centers of these galaxies cause a build-up of material around them that in turn powers tremendous releases of energy before the point-of-no-return.

R

Radiation The emission by a substance of energetic waves or particles, especially as part of a process of nuclear decay. While there are many safe forms of radiation (our bodies radiate thermal energy constantly!), it's still best not to put plutonium in your mouth.

Rainbow A brightly colored arch-shaped optical effect seen in the sky under certain conditions. Caused when sunlight is refracted and dispersed by water droplets or ice crystals in the air. Known to attract smiling clouds and unicorns.

Reflection The action of throwing back light or other wave energy—rather than absorbing or transmitting it—by an opaque, solid surface such as a mirror.

Refraction The action of altering the direction of light as it is transmitted from one transparent medium through another of different density. A pencil resting in a glass of water looks bent because of refraction.

Relativity The relationship between physical laws and phenomena and the frames of reference in which they play out. Specifically, the observations of Galileo and Einstein that space, time, and motion are not absolute but dependent on one's point of view. On the subject of rocket science, Einstein's theory of general relativity greatly informs our understanding of orbital mechanics.

Reproduce To bring about offspring (essentially smaller copies of a plant or animal) through various means.

Rest A state in which an object is experiencing no accelerations or forces, though it may be under the influence of gravity. Such an object seems still to an observer in the same frame of reference. As with motion, there is no state of absolute rest but only rest relative to the local environment.

Retrieving The only reason we have *storing!* Retrieving means getting data that has been stored in a computer system.

Rocket A projectile, manned or unmanned, driven forward by reaction propulsion in accordance with Newton's third law (simply by forcing its exhaust backward, a rocket moves itself forward). Unlike other jet engines which need to intake air to function, a rocket engine contains all of its own fuel and oxidizers enabling it to operate in the vacuum of space.

Rocket Science A term often used as shorthand for extreme complexity for the purpose of contrasting something simpler ("washing the dishes ain't rocket science, y'know!"), the science of rocketry is, actually, quite dauntingly complex. Many of the other sciences in this book—including a great deal of engineering—have a role to play in sending us toward the stars! As such, rocket science represents a highly orchestrated group effort...but it's not brain surgery.

Roll Also called a *bank*, this is any rotational movement of a flying vehicle around the axis of motion (left wing goes down while right wing goes up or vice versa).

Roots The complex growth of branches and fibers beneath the soil that anchor a plant and absorb water and mineral nutrients.

Rules The physics of the large (Newtonian mechanics and general relativity) and the physics of the very small (quantum mechanics) each have established rules for how masses, energies, particles, and waves behave (although the quantum rules tend to get a bit fuzzy). These two sets of rules, operating at vastly different scales, are as of yet incompatible with each other.

S

Scale A reference frame of physical size or volume relative to other amounts. The scale of North America lies somewhere near the midpoint between atomic scale and the scale of the solar system.

Scattering The action of randomly spreading out light or other radiation through collisions with particles in the medium in which the light is traveling. Which is why it's wise to buy traveler's insurance.

Science Any of several organized fields of knowledge built up through careful observation and testing. Studying science makes one smart, hip, and extremely interesting. No, really—they've done studies (...these studies have, of course, been done by scientists)! ;-)

Skeleton The entire assembly of the bones in a body, giving it shape, support, and a certain amount of rigidity. Seen frequently in late October.

Skin The body's largest, heaviest organ (and the most visible!). Skin is the soft outer covering of our bodies. Made up of several distinct layers of tissue, skin serves to regulate temperature through perspiration (sweating) and to protect our internal tissues and organs. It also contains the mechanisms for the sense of touch, alerting us to even subtle changes in temperature and pressure!

Sky Broadly, the area between the top of your head and the limits of cosmological expansion. Some senses of the word refer only to the earth's atmosphere, however.

Slide-rule A simple, manual instrument used for performing quick logarithmic calculations.

Small The universe as experienced on microscopic scales, descending to subatomic realms so tiny that most common-sense can't fit inside! Particles and energies here often behave in unpredictable ways.

Software The non-physical, untouchable components of a computer, made up of machine-readable coded instructions called *programs*.

Solutions Chemical mixtures produced by dissolving one or more substances into another substance.

Space Not just "up there," but *everywhere!* Space refers to the physical volume of the universe—its length, breadth, and height.

Space-time Taken together as a whole, space and time represent the sum foundation of reality upon which all things and events—at any place and at any time—occur. Almost named "time-space," but space had a better agent.

Special Relativity Albert Einstein's game changing theory, published in 1905, which extended Galilean relativity—the notion that all physical laws apply equally under conditions of steady motion—to the measurement of light's velocity. This would yield a constant value for the speed of light regardless of how fast either the measurer or the light emitter was moving so long as those speeds were steady. The theory operates without respect to the effects of gravity, a factor that Einstein would tackle in his general relativity theory. Implicit in special relativity are non-intuitive consequences such as mass/energy equivalence and the flexible nature of space and time once thought to be absolute.

Addendum: Special relativity is actually even simpler than the nursery rhyme makes it seem. From Galileo on, it was a casually accepted fact that so long as an experimenter occupied a workspace that was not undergoing any changes in speed or direction (even if that workspace was moving very fast relative to its environment, such as a fast ship on calm waters or even a supersonic jet!) all experiments conducted therein would yield identical results to those performed at rest, i.e., on solid ground. All Einstein did was point out that Maxwell's equations, which calculate the speed of electromagnetic radiation in a vacuum, must be included in that old Galilean relativity. So no matter where you're doing the math—either in a fantastic spaceship cruising at nine tenths light speed or on boring ol' solid ground—you always get the same result when measuring the speed of light: 299,792,458 meters per second. Well isn't that special!

Speed of Light 983,571,056 feet per second, which is also 299,792,458 meters per second, which comes out to 186,282 miles per second or 670,616,629 miles per hour. Working out the square of *any* of these uses a lot of pencil.

Square[1] (Engineering) Not exactly the geometric shape but rather an instrument formed into either a T or L shape used to trace or measure ninety degrees or right angles.

Square[2] (Newtonian Physics/ $E=mc^2$) Also not the shape but rather the value of a number multiplied by itself. The square of the number 4, for example, would be shown as 4^2, representing 4×4 and having a value of 16.

Storing One of an information system's most basic functions, storing allows for the keeping of digitized data for future use. Such data is encoded onto recordable media such as magnetic tape, optical discs, or muffin. Did I say "muffin"? Sorry—skipped breakfast this morning....

Stressing The action of a force upon a material that would tend to deform it if the material was not strong or rigid enough. The wolf, with his huffing and puffing, was an expert stressor of construction materials, straw and sticks a specialty!

Strings One of the strongest candidates for a "theory of everything" is string theory—actually a collection of theories that advance the notion that reality, at its most fundamental, is made up of incredibly tiny vibrating strings whose vibrations "sing" into shape all particles, all masses, and all energies. Bravo!

Stuff Sometimes called *toe jam*, this grey, squishy substance is a mixture of sweat, glandular oil, dirt, and sock lint. If you don't wear socks then it's just sweat, glandular oil, and lots of dirt.

Subatomic A hand is smaller than a person. A finger is smaller than a hand. A fingernail is smaller than a finger. A cell is smaller than a fingernail. A molecule is smaller than a cell. An atom is smaller than a molecule. Anything smaller than *that*...is subatomic (and difficult to count!).

Sun Did you know that the sun is a star? Well, it *is!* It only looks much brighter because it's so much closer than other stars. Being about 93,000,000 miles away, having a diameter of over 865,000 miles across and commanding over 99 percent of the solar system's mass, our very own star is the most noticeable object in the daytime sky! Try not to notice it directly with your eyes, though—its brightness can damage them!

System A group of organs and associated tissues which collectively carry out a single process in a body. The cardiovascular system, for example, circulates blood and oxygen throughout the body and involves the heart, arteries, veins, and blood vessels.

T

Telescope An optical instrument containing lenses or mirrors that manipulates light in such a way that distant objects appear to be much closer. While we may never know who first looked through two pieces of shaped glass to magnify images, most historians agree that the telescope was invented sometime in the sixteenth or seventeenth century by either Englishman Leonard Digges, German-Dutchman Hans Lippershey, or one of their contemporaries. No one doubts, however, that it was Italian super-scientist Galileo Galilei who first popularized the telescope as an instrument for astronomical observation after adding a number of his own improvements to earlier designs.

Tendrils The thin anchoring organ of ivies and other climbing plants. Tendrils can grow as fast as a foot a day!

Theory of Everything (ToE) Often called unified theory, this is a hypothetical theory that merges general relativity and quantum mechanics into a single set of concise equations expressing how the whole of creation works at all scales. We may need a lot of chalk.

Thermometer Any instrument that mechanically or electronically records and displays the local air temperature. Or mouth temperature if you're not feeling well!

Thrust The force that propels a vehicle forward produced by an aircraft or rocket engine. Automobile engines don't produce thrust, per se, but rather rotational energy which turns the wheels!

Time The stuff that keeps everything from happening all at once! Time is, in fact, one of the four dimensions of our reality. Along with length (back and forth), height (up and down), and depth (near and far), time forms the invisible background structure of creation called space-time upon which all things and events in the universe occur. Our experience of life moves constantly along a one-directional axis of time from past to future. This is technically time travel but, as it takes an hour to go forward an hour, it's nothing to get excited about.

Tissue A grouping of similar cells arranged to perform a particular function within an organ or wider body.

Transmission The movement of light through a transparent medium with little or no change in direction.

Trees Typically large, long-lived woody plants with a single stem—called a trunk—supporting outgrowing branches well clear of the ground.

U

Universe All of everything everywhere! Current science teaches that the universe is about 13.8 billion years old and about 10 billion light-years across. The uni- part of the word seems to indicate that there's only one of these things. Evil, bearded Spock tends to disagree.

V

Veins Another important component of the circulatory system, veins (along with arteries, capillaries, and blood vessels) form the highways, streets, and lanes by which blood is pumped throughout our bodies. Specifically, veins are the tubelike vessels—usually close to the skin's surface—through which deoxygenated blood travels from the tissues back toward the heart.

Violet A flowering plant that favors mild climates with five-petalled blooms that tend to be purplish or bluish in color.

Visibility An estimation of the distance a person of average sight can see as influenced by fog, precipitation, cloud cover, pollutants, and light level. If none of these factors are a problem, visibility is commonly labeled unlimited, which doesn't mean you can make out what color cufflinks are in fashion on the second planet of the star Epsilon Eridani.

W

Walls Plant cells differ most noticeably from animal cells by the presence of cell walls—layers of tough material surrounding the plant cell's outer membrane. Cell walls serve as protection and support as well as filtering substances passing into or out of the cell.

Water A colorless and odorless liquid formed as a compound of hydrogen and oxygen, chemical formula: H_2O. Water is the basis of all fluids in plants and animals. And soft drinks.

Wave A regular disturbance in an electromagnetic field that acts to transfer energy as it moves outward from its source.

Waveform A less rigorous stand-in for wave function. This refers to a measure of probability concerning the qualities of a given particle—its position, momentum, whether or not it cut itself shaving that morning and the like. An un-collapsed wave function indicates an unobserved particle, free to exist simultaneously in a number of states.

Waves Plural of wave—regular variations in electromagnetic fields, such as light, as they move outward from their source.

Weather The condition and action of the atmosphere at a specific place and time involving temperature, wind, air pressure, humidity, rain or snowfall, etc.

Weather Balloons Ranging in size from three feet to over twenty feet in diameter, these instrument platforms allow for measuring temperature, pressure, humidity, and more at altitudes of up to twenty-five miles!

Weight A measure of the amount of force being exerted on an object's mass by a gravitational field. The weight of an object can vary greatly depending on the strength of local gravity.

White light Visible light in its entirety. When all wavelengths of light are emitted equally they appear neutrally colored or "white" to our eyes. Within every beam of white light, however, exists every color of the rainbow as well as many more our eyes can't see!

Willow A type of tree found in cold to mild climates and usually near water typically having narrow, serrated leaves and tough roots.

Wings The wings of an aircraft have a very special shape along their cross-section called an airfoil (or "aerofoil" if you're British!). This shape, curved longer and higher on the top surface of the wing, causes air to move over the wing more quickly than it moves under it. This makes for higher pressure under the wing which in turn creates lift.

X

X-Ray a type of short wavelength electromagnetic radiation above the visible range, capable of passing through solid materials in exactly the way that visible light doesn't.

Y

Yaw Any side-to-side rotational movement of a vehicle (nose goes left while tail goes right or vice versa).

Z

Zinc A metallic element, silvery or bluish-white in color, used extensively in forming alloys and galvanizing other metals to prevent corrosion.

About the Author

JOHN M. MULHERN is a fancier of science, silliness and taco Tuesdays at least once a week. Raised in Pawtucket, he works as a graphic designer in Rhode Island. He and his partner Sue (who shares those first two passions) enjoy astronomy in particular, science fiction in general and long walks on the beach when they can't be avoided. It's the author's fervent wish that this book—beautifully illustrated by Elisha Gillette—helps to cultivate an interest in the sciences among young people everywhere. Or, at the least, sells well enough to defray his mounting not-at-all-fake moon rock debt.

About the Illustrator

ELISHA GILLETTE learned to draw by watching a lot of behind-the-scenes features on their favorite DVDs. They have called Providence home since graduating from the Rhode Island School of Design in 2021, though they miss their family back in San Mateo, California. They are currently having a blast making art for books, independent clients, and fellow nerds. When not drawing, Elisha is most often observed playing tabletop games, untangling yarn, cooking too much for one person, or staring at birds. Elisha loves a good cup of chrysanthemum tea and meeing their neighbors.

www.elishagillette.art | elishagillette.carrd.co | elishagillette@gmail.com

www.ingramcontent.com/pod-product-compliance
Lightning Source LLC
Chambersburg PA
CBHW041323210926
51847CB00002B/29
* 9 7 8 1 9 6 3 2 9 6 2 8 0 *